An Early Start to Mathematics

Roy Richards

Lesley Jones

SIMON & SCHUSTER

LONDON • SYDNEY • NEW YORK • TOKYO • SINGAPORE • TORONTO

Text © Roy Richards and Lesley Jones 1990
Design and artwork © Simon & Schuster 1990

First published in Great Britain in 1990 by
Simon & Schuster Ltd
Wolsey House, Wolsey Road
Hemel Hempstead HP2 4SS

Reprinted 1990

Printed in Great Britain by
BPCC Paulton Books Ltd

British Library Cataloguing in Publication Data

Richards, Roy
    An early start to mathematics
    1. Mathematics
    I. Title   II. Jones, Lesley
    510
    ISBN 0–7501–0034–6

Series editor: John Day
Editor: Elizabeth Clarke
Design and artwork: David Bryant/Joan Farmer
                    Artists

The importance of establishing a sound mathematical base in the infant school is widely recognized. The development of concepts, skills and attitudes are best planned through the four strands which can be isolated. These form the four main sections to this book: algebra, number, measurement and shape. Data handling, the fifth section, is briefly touched upon and spans all four strands.

Many of children's everyday decisions, games and problems require a logical approach. Such logical thinking is essentially a mathematical activity and one which provides a necessary foundation for an understanding of number. Measuring involves estimation, approximation and gradually developing the ability to use suitable measuring devices with some accuracy. Shape, in its three dimensional aspects, is a part of mathematical activity which involves children in exploring their world.

An underlying philosophy of the book is that mathematics should emerge from practical activities which are relevant to the child. Many of the ideas illustrated here depict everyday events in school and we attempt simply to make explicit the mathematics embedded in them.

Language has an important role to play in mathematical development. A child can develop language ability through the provision of suitable practical activities and through interaction with his/her teacher and other children. At the same time, this interplay can significantly improve the child's cognitive understanding of mathematical ideas. Children will be involved in listening to explanations, forming questions and answers, explaining and discussing problems. They should be encouraged to observe and look for patterns and relationships as well as practising skills which are important. There is a variety of apparatus which will provide a stimulus for such work. This includes the microcomputer and the calculator, both of which are increasingly available in the primary classroom and at home.

This book gives a concise, comprehensive introduction to the mathematics that young children should cover.

Make some sound patterns

Clap some patterns. Put some 'pops' in between the claps.

Allow children to invent simple patterns for others to follow.
Musical instruments can also be used to produce repeated patterns.

maracas   tambourine

Make some visual patterns

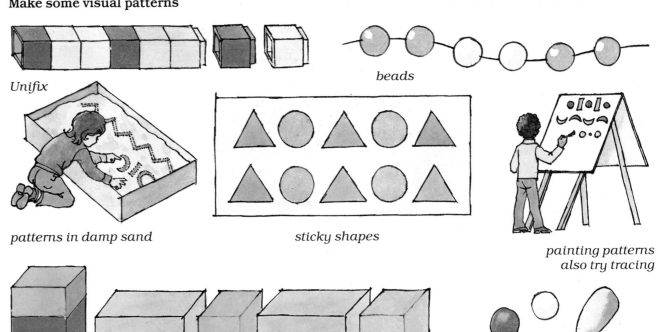

Unifix

beads

patterns in damp sand   sticky shapes

painting patterns
also try tracing

building bricks

pebbles

party hats

Playpeople

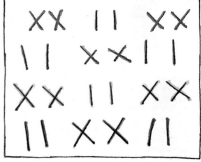

sewing patterns

Talk about the patterns made by the children.

Make some North American Indian patterns

Play 'Follow my leader'. The leader sets the pattern for the others to follow.

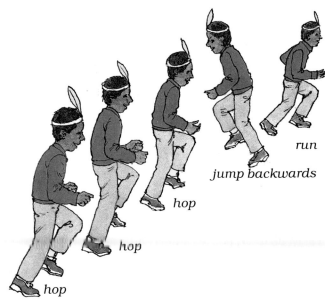

run

jump backwards

hop

hop

hop

Play some pattern games

Play team games where the activities follow a set pattern.

run along a bench   *climb in a box*   *go through a hoop*   *jump over a mat*

Racing games

Racing games help with counting, matching one to one, and with order.

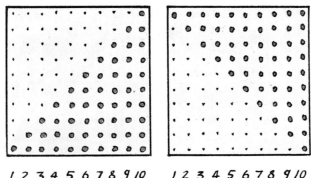

Which horse finishes first, second, third?

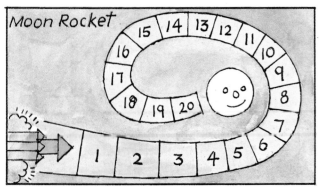

Pegboard patterns

These patterns can display order.

1 2 3 4 5 6 7 8 9 10   *1 2 3 4 5 6 7 8 9 10*

Obstacle races

Run between the hoops.
Who finishes first, second, third, last?

Fill in the missing sets

Building activities

Which brick did you use first, second, third and so on?

Make a chart to illustrate these sets.

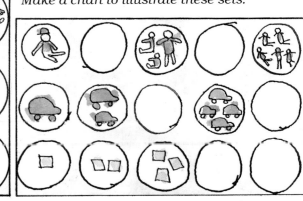

Big dipper game

Match your card with a number on a car on the big dipper.

Stepping stones

Step across the carpet tile 'stones' in order.

Step in the hoops in order.

Stories emphasizing order

Children or parents may be able to provide stories from different cultures.

Use picture cards

Discuss sequences relevant to the children's own experience: for example, getting up, how a plant grows, or bonfire night. After discussion, use picture cards to show the sequence. Jumble up the sequence and let the children put the cards in order.

Number rhymes

Three little speckled frogs
Sat on a speckled log,
Catching some most delicious bugs,
yum, yum.

One jumped into the pool,
Where it was nice and cool,
And then there were two green speckled frogs,
glub, glub.

Two little speckled frogs.

Story time

Use familiar stories to reinforce number work.

Dice games

Any game where the number rolled on the die has to be translated into a move reinforces the idea of the cardinal aspect of number. Use dice which are marked in non-conventional patterns, or spinners which introduce symbols.

Introduce the concept of zero

How many children are wearing red shoes?

None.

How many people have an elephant as a pet?

None.

Number game

Ask your partner for the number of counters shown by the symbol. Then check that they cover the dots on the card.

Cards 1–9

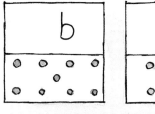
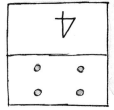

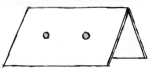

fold card so that it stands up

Early activities leading to pictorial representation

Use objects to represent people.

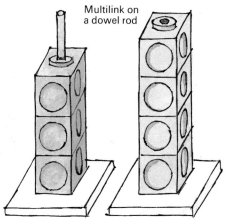

Multilink on a dowel rod

3 children are wearing green jumpers.

4 children are wearing blue jumpers.

6 children are wearing trousers.

3 children are wearing skirts.

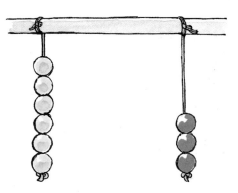

Play games which introduce the idea of a symbol representing a number of things.

Counting sheep

See if you can put the same number of sheep in your pen

Use symbols for instruction in PE

Do 3 bunny hops

Do 2 rolls

Do 6 jumps on the spot

Playground symbols

Run and make the right number of children on each symbol.

Put the right number of toys in each truck

Make the correct number of chimes

Clothes shops

The shop assistant has to give a number symbol to correspond with the number of items chosen for trying on.

Now you see it – now you don't

Use the beakers to cover some counters. Let the children use symbols to show how many counters they think are hidden.

Postman's knock

Show the number symbol to correspond with the number of knocks on the door.

Play games that introduce grouping and exchanging.

Fishing

Put three fish in a net.
Put five fish in a net.

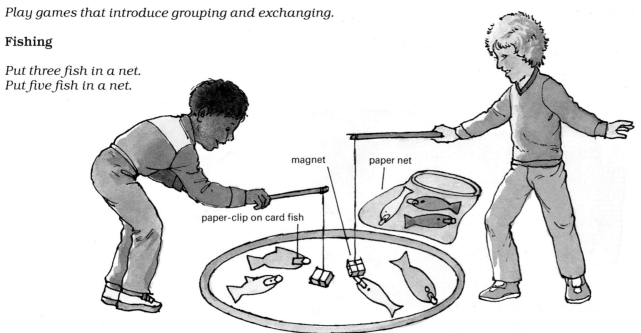

magnet

paper net

paper-clip on card fish

Dice games

Use different number bases.

Collect the number of units shown on the dice.
Exchange five units for one rod.
Exchange five rods for one flat.
Exchange five flats for one cube.
The first to a cube wins.
Working with two dice speeds up the game and
allows the children to deal with larger numbers.

Dienes Multibase Arithmetic Blocks

Use units, longs and flats.

Pick a card and make that number.

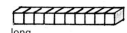

unit

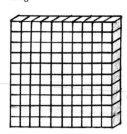

long

flat

pack of numbers

place mat

12

110

17

Calculators can help

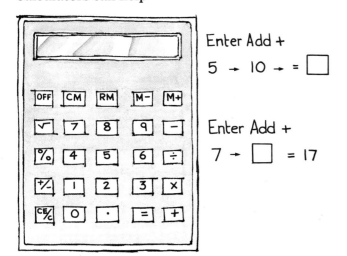

Enter Add +

5 → 10 → = ☐

Enter Add +

7 → ☐ = 17

Computer programs can help

Spike abacus

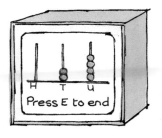

H   T   U
Press E to end

Counter

12

Microprimer disc

Slimwam disc

The number-line can help

Count on ten from there.

Grouping leading to place value

Take a collection of items: for example, seven pencils.

Group them in different sets. Complete sets move to the left. Make a table of results.

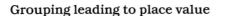

Sets of 2	Ones
Sets of 3	Ones
Sets of 4	Ones

Lucky tens game

Moving in tens helps with place value.
Use a die to move.
When you land on
a horseshoe, you
move on ten.
On a banana skin,
you slip back ten.

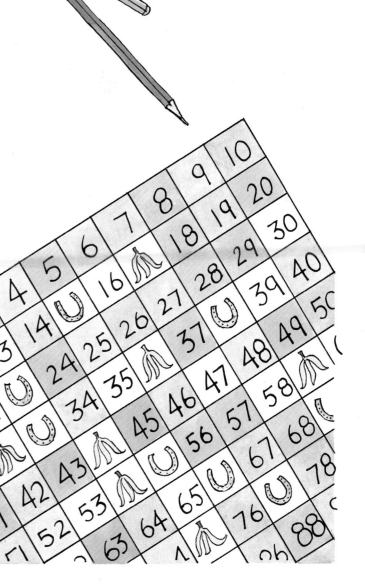

Multilink, Cuisenaire, abacus

Continue the pattern.

Cuisenaire

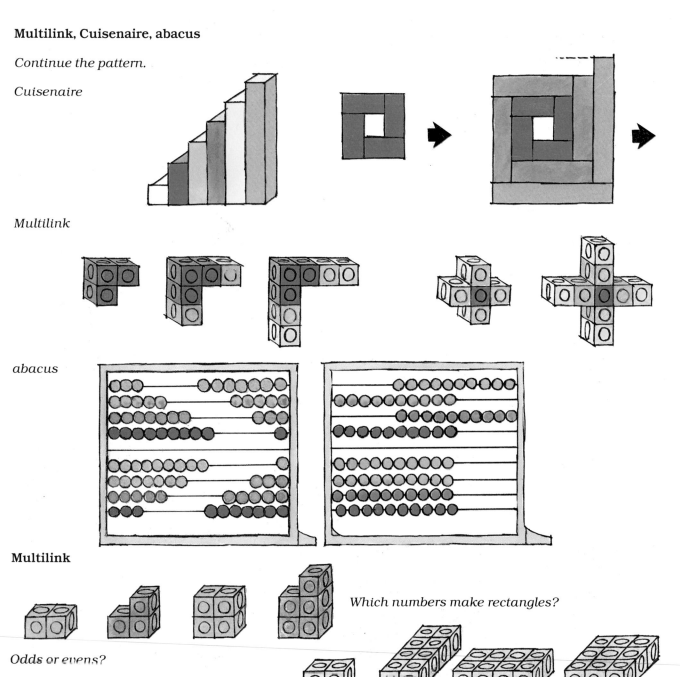

Multilink

abacus

Multilink

Which numbers make rectangles?

Odds or evens?

Cuisenaire rods

How many different walls can you build?

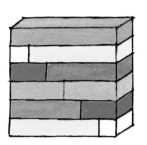

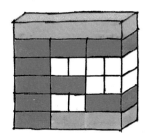

Number-line

Use the number-line for: addition
subtraction
multiplication.

Using Lego

Make a collection of Lego in these sizes:

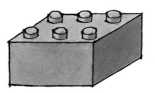 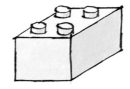

How many different walls can you make, using this brick as a base?

 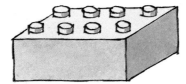

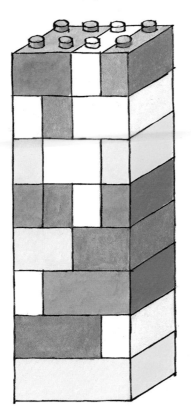

Arranging beads

How many ways can you arrange three white and two black beads?

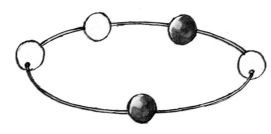

Colouring flags

How many different flags can you design using red, blue and green?

Using stories

How many ways can Goldilocks and the three bears sit down at this table?

Partitioning sets

Where a set is already clearly defined (for example, a set of all the building blocks), we can partition the set.

These will not roll.   *These will roll.*

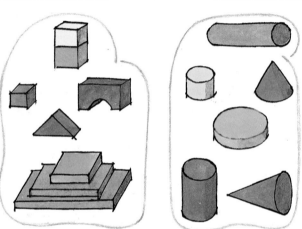

Tidying

Tidying activities can include sorting.

Sorting activities

These help to develop and refine language.
Sort on: colour – black, grey, red etc.
texture – rough, smooth, pointed etc.

Sort on texture: rough
smooth
furry
silky
warm
cold
hard
soft

Sort blindfolded: bricks with every edge flat
bricks with some edges curved

Sorting into sets

Any criteria can be used for sorting but sets should always be well-defined so that you know what belongs and what does not.

This is the set of hats that I like.

Play sorting games

All those with red jumpers go home.

Let the children choose their own criteria. Initially the sorting should be into two sets. Draw attention to the negative aspects as well. If, for example, a child chooses 'All those children who are wearing jumpers', then the other set becomes 'All those children who are not wearing jumpers'.

Sorting by shape and colour

Stand up if you have the same colour as this but a different shape.

Separating red and green beads

Emphasize the negative aspects, too. For example, this is a set of beads which are not red.

Sorting by smell

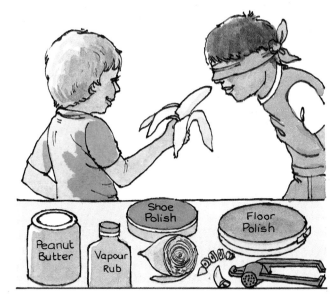

Peanut Butter · Vapour Rub · Shoe Polish · Floor Polish

Sorting by taste

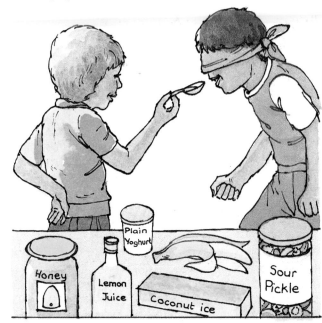

Honey · Lemon Juice · Plain Yoghurt · Coconut ice · Sour Pickle

Decision trees

Decision trees are symbolic ways of representing logical thinking. They are also used in everyday life.

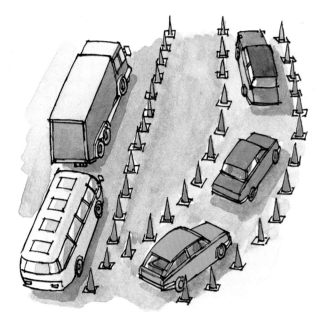

Sorting Logiblocs

Decision trees can be extended.

Make a wall display

Do decision-making on a large scale

black shoes   *not-black shoes*

Decision game

What is the criterion for sorting?
One child makes the decision.
Then makes up the cards.
Then hides the cards.
Friends look at the sorting tree to decide what is drawn on the hidden cards.

Ways of representing data

tall   short

Carroll diagrams

Venn diagrams

Play 'What's in the square?'

Clothes combinations

Systematic ways of doing things are important.
These can be recorded on a matrix.

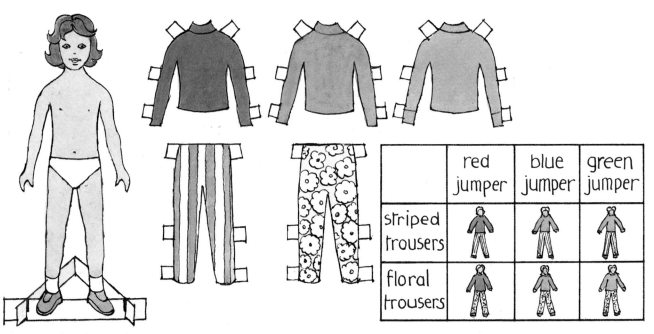

	red jumper	blue jumper	green jumper
striped trousers			
floral trousers			

Progression in sorting

Progress from sorting one attribute ⟶ *to sorting two attributes* ⟶ *to sorting three attributes*

Use floor Logiblocs.

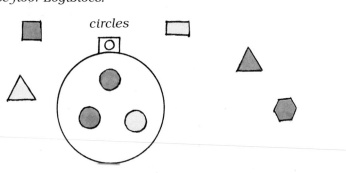

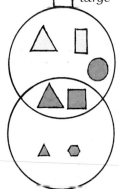

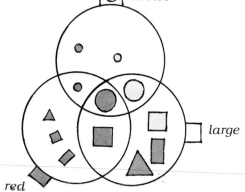

Using picture cards

Let children make their own systems for logic apparatus. For example, create some picture cards of teddies. Let the children add features systematically – an umbrella, a hat, different colours.

Jumble up the cards and sort them:

all green teddies

all teddies with hats and umbrellas

all blue teddies with umbrellas

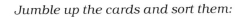

Matrix patterns

Use a matrix to create a pattern with your Logiblocs. Which Logiblocs are missing?

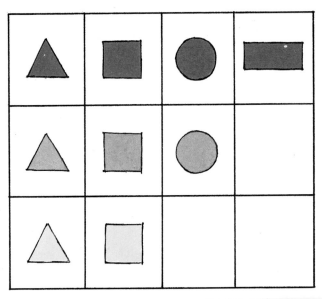

Arrange Logic People systematically.

'*Machines*' *provide an early introduction to change. This is basic to the mathematical idea of a function.*

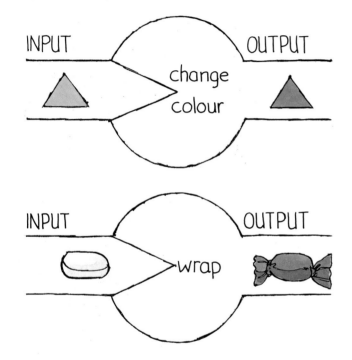

Two or more simple machines can be linked to produce a multiple machine. This has more challenge.

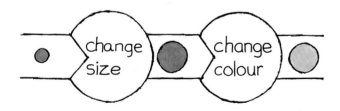

Machines like this are a good way of introducing the different kinds of relationships which exist between two objects.

Make a machine

The machine can be operated from within or without.

Drama, art and stories

The machine theme can be developed through drama, art and story-telling.

Try simulating piston action.

Relationships between two objects

Jill fell in the chocolate machine.

The Mars bar that fell into the sausage machine.

Children will need to use logic to play these games.

Password game

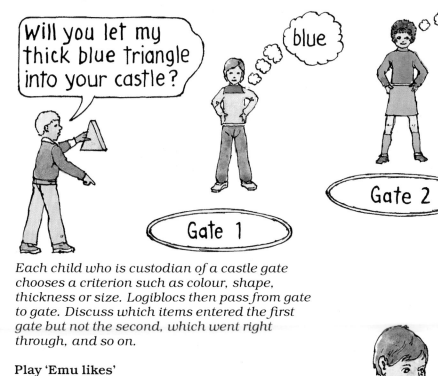

"Will you let my thick blue triangle into your castle?"

blue

thin

circles

Gate 1

Gate 2

Gate 3

Each child who is custodian of a castle gate chooses a criterion such as colour, shape, thickness or size. Logiblocs then pass from gate to gate. Discuss which items entered the first gate but not the second, which went right through, and so on.

Play 'Emu likes'

Emu likes horses but he doesn't like carrots. Emu likes dogs but he doesn't like bones. Emu likes sheep but he doesn't like wool. Do you think Emu likes goats?

Likes Dislikes

Make a picture book to show the things Emu likes and dislikes.

Discussion helps

If it is Sunday then I will not

Watch Blue Peter

Get up early

Go to school

Many of the ideas in logic are dependent on precise use of language. Children need to understand the use of relationships which include 'not', 'if...then', 'and...or' and 'neither...nor'. These can be developed through discussions and games.

Domino game

Change one attribute each time (size, shape, thickness or colour).

What are the rules?

One child sorts according to given rules. The other children guess what the rules are.

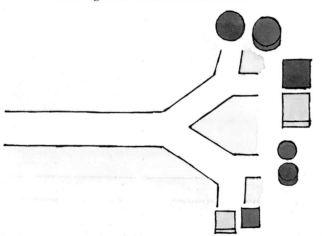

Making a pattern

One child makes a pattern, then instructs another how to make it. When the pattern is completed the screen is removed.

Which attribute?

One child thinks of an attribute and picks out all the shapes with that attribute. Other children have to guess the chosen attribute.

Elephant's bun feast

Make the Jumbo Mansion from eight shoe boxes. Use home-made dough for buns.

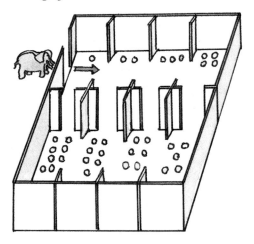

How many buns can the elephant eat without entering any room more than once?
How many different solutions?

Traffic wardens

A traffic warden can see the length of one block in any direction. What would be the minimum number needed to patrol the area shown and where would you place them?

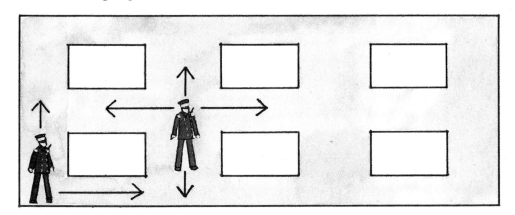

Building bungalows

Use five Multilink cubes for each. How many different designs can you make? Children may need to turn shapes round and fit them together to decide which shapes are identical.

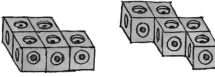

Playing with square numbers

Use red or blue Multilink cubes or pegboard pegs. Which square numbers use equal numbers of each colour?

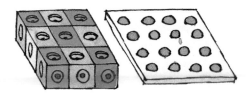

Just patterns

How many different patterns can you make with these counters?

Going home

How many different routes from home to school?

Shongo network patterns

Children in Zaire can create patterns like these without lifting a finger or going over the same ground twice. Can you?

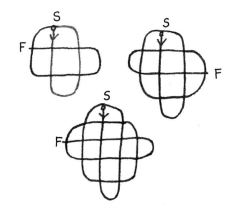

Picture moving

Arrange the pictures in the grid so that no two identical animals are in the same row or column.

grid for pictures

Number has no concrete existence: it is always a property of things. Number is a property which refers to sets of objects. There is no single object that can have the property 'two', but a set of objects can. So before studying number, sets of objects must be studied. Combining and partitioning sets helps.

Using Multilink

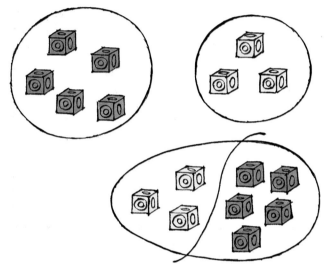

Using Cuisenaire rods

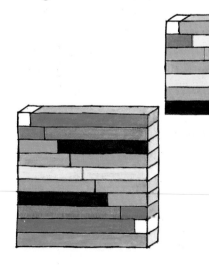

(1,6)
(2,5)
(3,4)
(4,3)
(5,2)
(6,1)
(7,0)

Discussion helps

Encourage the children to talk about number situations.

Six of the yachts are sailing. Two have capsized. There are eight altogether.

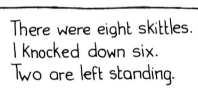

There were eight skittles. I knocked down six. Two are left standing.

Fairground balloons

Join the pairs which make ten.
Change the balloons for new number pairs.

Party-time

Ask children to choose a total of six items for party-time.

Use the string to match the number of cakes with the number of samosas to make your total of six.

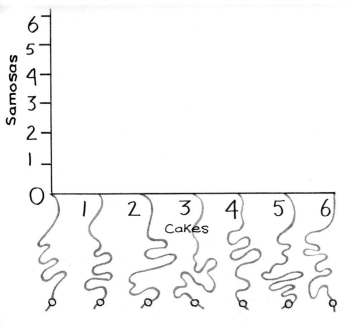

Addition squares

Look for patterns in the addition squares. Where do the sixes occur?

Section off a square within the main square. What patterns can you see?

+	1	2	3	4	5	6
1	2	3	4	5	6	7
2	3	4	5	6	7	8
3	4	5	6	7	8	9
4	5	6	7	8	9	10
5	6	7	8	9	10	11
6	7	8	9	10	11	12

Using tabards

Join together to make 10.

Join hands with a number 2 greater than yours.

All the odd numbers clap.

All the even numbers hold hands.

All those less than 7 make a circle.

Things to measure with

Make a collection of things to use for measuring.

Things for sorting

Make some collections for sorting.

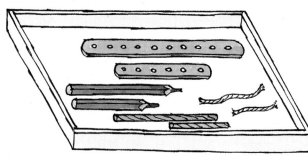

long and short box

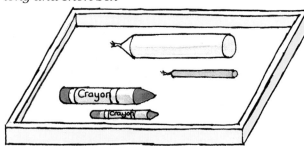

thick (fat) and thin box

wide and narrow box

tall and short box

Comparison of two unequal lengths

Children find it easier to begin with differences in linear measurement.

Find things longer than a pencil.
Draw the pencil on a piece of card.

Find things shorter than a ruler.
Draw the ruler on a piece of card.

Vocabulary

long, short, longer than, shorter than, tall, small, wide, narrow, thick, thin.

Matching lengths

Make up a box containing 'pairs' of objects of about the same length.

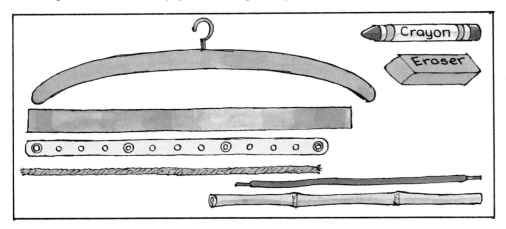
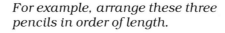
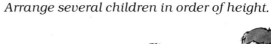

Jumble them up and let the children do the matching.

Measuring lengths

Use several objects to measure a length.

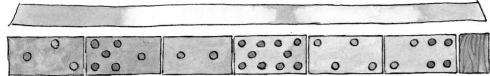

This scarf is about the same length as four pencils and a paintbrush.

This ribbon is about the same length as six dominoes and a brick.

This shoe is about the same length as four rubbers.

Use such arbitrary units as these for an exercise with real purpose. For example, will a table fit into a space between two cupboards?

Ordering

Once children have a grasp of the terms 'shorter than', 'longer than', 'taller than', they can move on to ordering things.

For example, arrange these three pencils in order of length.

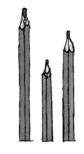
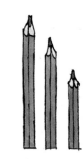

Arrange several children in order of height.

Use Dinky toys for ordering

In order of height:

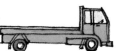

In order of length:

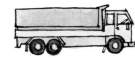

Vocabulary

longest, tallest, shortest, as long as, as tall as.

Some information

The old measurements of cubit, yard and fathom are interesting. For the true values of these measures one should use an adult. Nevertheless, it is fun and mathematically valuable for children to use their own body parts to measure length.

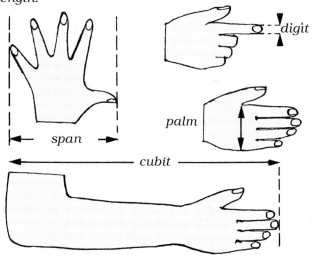

span

palm

digit

cubit

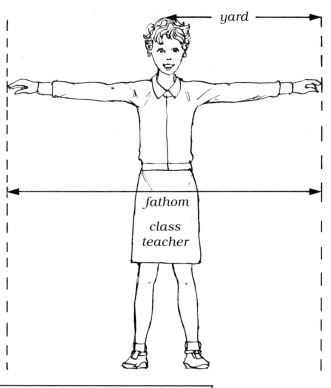

yard

fathom

class teacher

Span	Distance from the tip of the little finger to the tip of the thumb when the hand is outstretched.
Cubit (two spans)	Two spans make one cubit. Distance from the tip of the longest finger to the elbow.
Yard	Two cubits make one yard. This relates back to the practice of measuring cloth from the nose to the fingertip.
Fathom	Two yards make one fathom. This is very approximately equal to the height of a grown person.

Cut-outs

Cut out handspans, footprints and cubits.

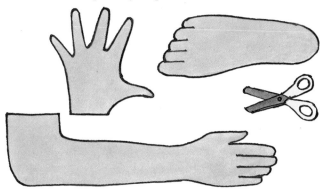

Use them to measure the length of objects. Measure a table in spans and cubits.

Measure a cupboard in spans and cubits.

Look for the relationships between measurements in spans and cubits

Estimating length

How many feet from the hall to the classroom? Guess first, then measure.

How many children can lie head-to-feet in the corridor? Guess first, then measure.

How many digits long is the bookshelf?

How many fathoms wide is: the playground? the school hall?

Make a class-span measure

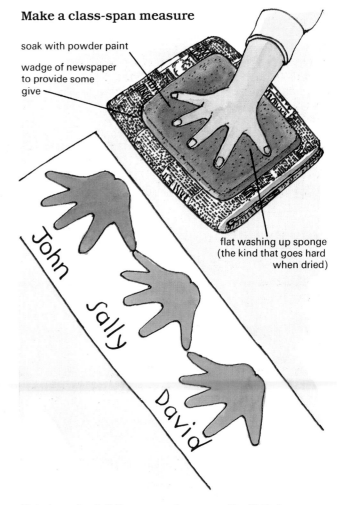

soak with powder paint

wadge of newspaper to provide some give

flat washing up sponge (the kind that goes hard when dried)

John Sally David

Print each child's span along a roll of kitchen paper. Make sure that they just touch and include every child in the class.

Use your class-span measure to find the length of: the school hall the playground the street.

Work like this, which uses parts of the body as measuring devices, helps children to realize the need for standard measures.

Samuel I Chapter 17 Verse 5

'Goliath, of Gath, whose height was six cubits and a span.'

Make some cut-out figures of Goliath, one to mount on the wall and one to hang freely. You need to use adult cubits and spans to match the biblical measurements. Measure the tallest teacher in the school.

How do the teachers in the school match up with Goliath?

Can Goliath fit on the classroom wall or must he be mounted in the school hall?
Is Goliath more than three times the children's average height?

Fix string to an eye-hook in the centre of the wooden batten. Tape string to the back of the head to prevent it from lolling forward.

Genesis Chapter 6 Verse 15

'The length of the ark shall be three hundred cubits, the breadth of it fifty cubits, and the height of it thirty cubits.'

If you have a large enough playground, you can draw the ark full size.

Compare different parts of the ark with parts of the school building.
For example: the width of the ark with the length of the school hall.
the height of the ark with the width of the school hall.

You could make a wall-mounted model in sugar paper. You would need to work to scale. The children will probably not understand the scale but you could make the model to the size given below to get an appreciation of its proportions. A scale of 1:20 is feasible: that is, make the length 15 cubits and the height 1.5 cubits. Even so, this would probably need to be hung on the wall of the school hall. The model will give a very good idea of the ark's proportions.

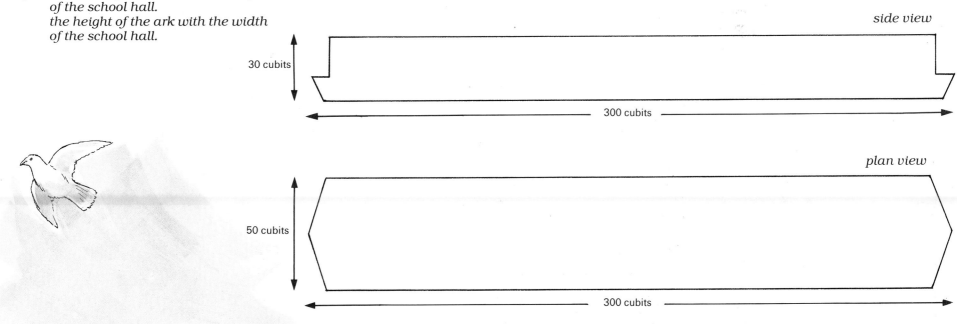

side view

30 cubits

300 cubits

plan view

50 cubits

300 cubits

1½ cubits

15 cubits

Work with standard units needs to be prefaced by lots of work with arbitrary units to establish the need for standards.

Things needed

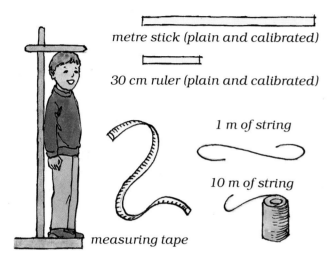

metre stick (plain and calibrated)

30 cm ruler (plain and calibrated)

1 m of string

10 m of string

measuring tape

Start with a plain metre stick

Find objects longer than, shorter than, and about the same length as a metre.

about 1 m   longer than 1 m

about 1 m

shorter than 1 m

less than 1 m
less than 1 m

handspans in a metre

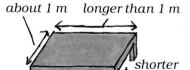

my span is about a metre

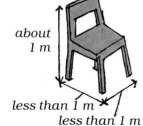

feet in a metre

cubits in a metre

Estimate – then measure with a metre stick

How long is the corridor?

How wide is the school hall?

How high is the classroom door?

Estimate – then measure with a 10 m string

How long is the playground?

How long is the street?

You might want to add on length measured with a metre stick to get a more correct length.

Vocabulary

About one metre, nearly three metres, just over three metres.
Half metres can be judged also, so use phrases such as 'about one and a half metres'.

Work with a decimetre rod

Measure small items. Check how many decimetres make a metre.

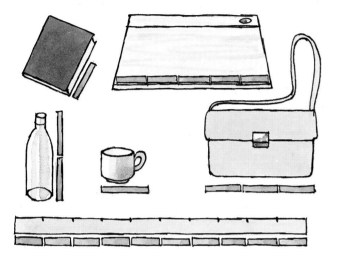

Work with centimetres

Measure lines of different length drawn on paper.
Measure objects and keep a record.

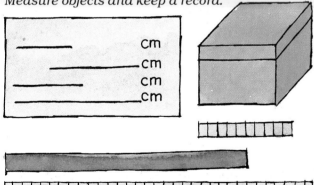

cm
cm
cm
cm

Measure curved lines by using string.

Measure yourself

Cut out strips of paper to correspond to the height of each child.
Make a chart of heights.

Our heights

Roy John Rupee Tim Pat Jill

span   foot

waist

head circumference   pace

Draw up a table for the class.
Who has the largest handspan? The largest foot? The largest pace?

Are some handspans, feet or paces the same length as others in the class? Is there a link between height and foot length?

Make a model from card

Make your model life-size.

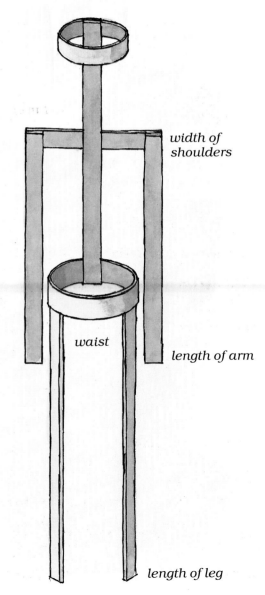

head circumference

width of shoulders

waist

length of arm

length of leg

Children are well aware of the phenomenon of weightlessness through hearing about the work of astronauts. It is therefore necessary to establish the concepts of mass and weight at an early age.

The mass of an object is the amount of stuff in it. Mass is measured in grams (g) and kilograms (kg).

If a person is placed in an equal pan balance on the earth, and then on the moon, the balance will be the same in each case.

In this instance, masses are being compared. The mass of the person on the moon is the same as his/her mass on the earth.

The weight of an object is the force exerted on it by a gravitational field. Weight is measured in newtons (N).

If a person were to stand on bathroom scales on the earth, and then on the moon, the readings would be different.

This is because the moon's gravitational pull is about one-sixth that of the earth. The person would therefore weigh on the moon about one-sixth of what he/she weighs on the earth. Here weights, not masses, are being compared.

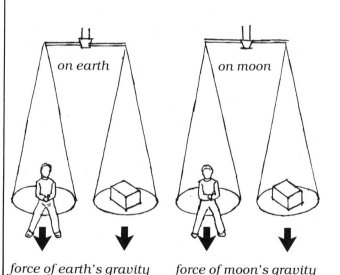

on earth   on moon

force of earth's gravity   force of moon's gravity

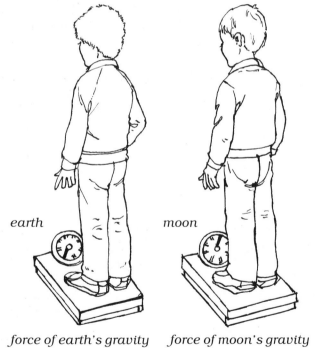

earth   moon

force of earth's gravity   force of moon's gravity

Vocabulary

heavy/light   heavier than/lighter than

has greater mass than/has less mass than.

Heavier than/lighter than

Explore situations where these terms can be used.

The wooden ball is heavier than the tennis ball. The wooden cube is lighter than the marbles.

heavier than my pencil box

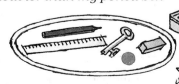

lighter than my pencil box

Getting a balance

Use matched pairs of objects.

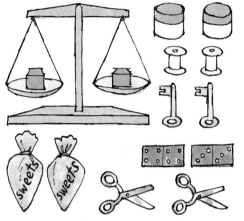

Now try balancing one thing against sets of other things. First estimate, then try.

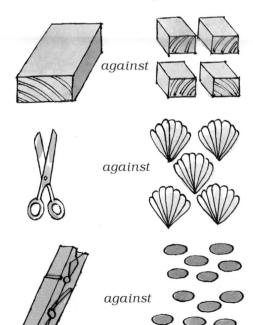

against

against

against

Ordering

is heavier than   *is lighter than*

is heavier than   *is lighter than*

Use a balance to help

Susan's shoe balances six blocks. William's shoe balances seven blocks. Jane's shoe balances nine blocks.

Try ordering parcels which vary in size, shape and mass.

Standard units

A 100 gram mass is easiest to work with. Make some up by putting sand in matchboxes.

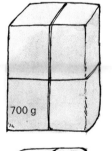

Weigh peas, beans, marbles, pebbles, shells, Lego, conkers, buttons, bottletops, counters and beads. Weigh all these in units of 100 grams.

Weigh two half-kilos against one kilo.
Find things lighter than a half-kilo, such as a shoe.
Find things heavier than a half-kilo, such as a brick.
Weigh a kilo of sand, pebbles or fruit.

Mystery parcels

Can children put them in order by lifting them to 'feel' how heavy they are?
Add a parcel of 500 g mass. Can children pick it out?
Put in parcels greater than 1 kg and less than 100 g.
How easy are these to find?

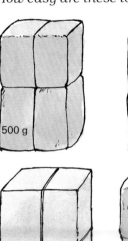

500 g

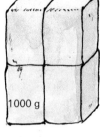

1000 g

700 g

500 g

300 g

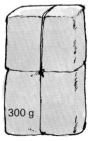

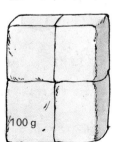

100 g

Vocabulary

balance, heaviest, lightest has greater mass than/has less mass than.

Capacity: the amount of space in a hollow container (ml and l)
Volume: the amount of space occupied by a solid shape (cm³ and m³).

Water and sand-play supply a rich vocabulary

full/empty
more/less
how much more/how much less
holds more/holds less.

Capacity

Five cups fill one teapot.

Six teapots fill one bucket.

Making sets according to capacity

These hold more than the bottle.   *These hold less than the bottle.*

Cubic decimetre

Water volume is 10 cm × 10 cm × 10 cm = 1000 cm³
Capacity of container is one litre = 1000 ml
Mass of water is one kilogram = 1000 g

So a cubic decimetre has an edge of 10 cm, a volume of 1000 cm³, fills a container of capacity 1000 ml and has a mass of 1000 g.

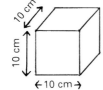

Note: 1 cm³ = 1 ml = volume filled by 1 g of water.

Standard measures for capacity

Use these for measuring with.

litre jug   litre carafe

litre bottles

litre carton

Measuring capacity

How many of each of these in a litre?

How many litres in each of these?

With experience, how well can children estimate capacity?

Measuring volume

Examine the volume of solids.

Using cubes, construct a variety of 'solids' that will occupy the same volume. Here are some examples using 12 cubes.

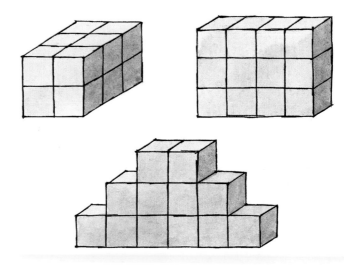

Make some boxes to hold the cubes.

These differently shaped boxes have the same capacity.

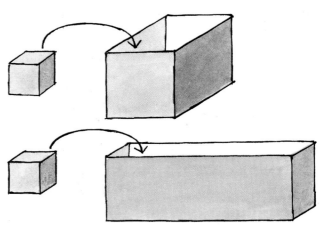

There are two important concepts to develop: the sequence of events and the duration of events. Here are some activities which will help to develop an idea of the sequence of events.

Draw some pictures

Things done during the day.

Things done or happening at night.

Pictures can be wall-mounted or put in a concertina folder.

Make a picture clock

Lunch
PE
Playtime
Going to school
Painting
Getting up
Getting ready for bed
Going home

Vocabulary

before, after, yesterday, today, tomorrow, night, day, first, next, earlier, same time as.

Put the picture cards in order

Getting up

Breakfast

Going to school

Assembly

Milktime

Lunch

Make some sequence cards – put them in order

Begin with a comparison of things done quickly against things done slowly.

Race across the playground

Who is the fastest?
Who is the slowest?

Dressing after PE

Who is the quickest?
Who is the slowest?

Bouncing a ball

Who can keep the ball bouncing longest?

Skipping

Who can skip for the longest time?

Make books illustrating the speed of movement

Use drawings, paintings and pictures cut from magazines.

My book of animals that move slowly

elephant, worm, tortoise

My book of animals that move quickly

cat, dog, horse, lion, deer

My book of vehicles that move slowly

road-roller, pram, milk-float

My book of vehicles that move quickly

planes, trains, cars, buses

Make some timers

water clock

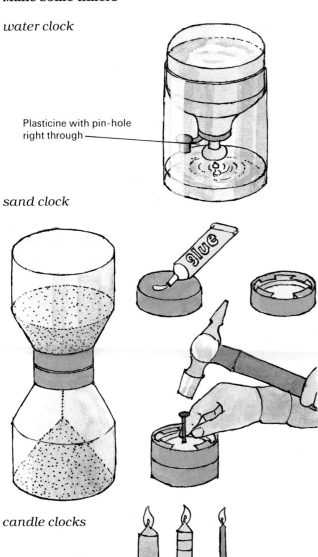

Plasticine with pin-hole right through

sand clock

candle clocks

One candle can be marked against the burning of another: for example, at 5 minute intervals.
Use birthday candles where you need quick results.
How far down do they burn during playtime?

Make a human sundial

On the playground, draw around a child's shoes first thing in the morning and draw the shadow. Draw the shadow at noon and again at the end of the day.

While the sand clock empties

How many bricks can you build into a tower?

How many beads can you thread?

How many 10p coins can you pick out of the box of money?

How many times can you write your whole name?

How many bounces of the ball?

Use a seconds timer

Time activities such as:
putting pegs in a pegboard
running around the edge of the school hall
hopping down a corridor.

All these activities give experience of the duration of events. Ordering the length of activities will lead to an eventual understanding of the need for standard units, such as the second and the minute.

Children need practical experience of measuring the duration of events in order to develop the concept of time. That they can read digital or analogue timepieces does not necessarily mean that they have this concept.

Digital timepieces are common. Use of these is easily taught and should precede teaching how to read an analogue timepiece.

1 Practise reading 1 to 59

Use the number-line.

Play number games, such as counting around in a circle.

Follow all the usual number work (see pages 4 to 14).

2 Practise the shape of numbers

Practise familiarity with the shape of the numbers on a digital timepiece.

Use Cuisenaire rods to make up the numbers.

3 Practise reading the hours

Concentrate on the two left-hand places.

Make 24-hour cards.

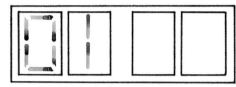

1 am

2 am

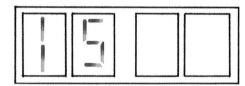

3 pm

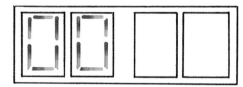

midnight

Play games to put them in order.

4 Practise reading the minutes

Concentrate on the two right-hand places.

Again, a series of cards for putting in order will help.

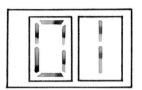

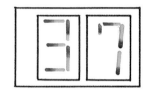

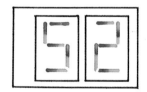

Show that

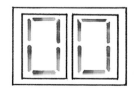

means 'o'clock'.

1 Times familiar to children

Begin with times familiar to children, even if these are 'awkward' times.

milk-time

home-time

Blue Peter

Let children play with a large clock face to familiarize themselves with the size of the hands.

2 Hours

*Now concentrate on the hours.
Fill in the missing hand.*

5 o'clock

3 o'clock

8 o'clock

3 Half-hours

*Fill in the missing hand/hands.
Long hand always points at the 6.*

half past 12

half past 4

half past 6

4 Quarters

quarter past 8

Long hand always points to the 3.

quarter to 5

Long hand always points to the 9.

5 Five minute intervals

Fill in the missing hand.

minutes past

minutes to

6 Practise conversions

5 minutes past 2 or 2 : 05

20 minutes past 4 or 4 : 20

25 minutes past 3 or 3 : 25

quarter past 7 or

half-past 1 or

5 minutes to six or

quarter to 10 or

20 minutes to 7 or

1-05 or 5 minutes past 1

4-20 or 20 minutes past 4

4-35 or

11-55 or

9-15 or

Sort money

Use real coins where possible.

Sort by colour.

Sort by value. Stick a coin with Blu-tack to the centre of each saucer.

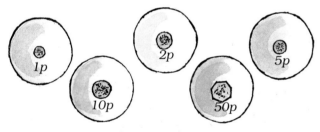

Now sort the coins into each saucer.

Sort by touch.

Shape, size, thickness, kinds of edges – all these give clues.

Coin rubbing

With children new to money you might concentrate on values up to 10p.
Make some rubbings of the different coins and mount them on paper.

Order the coins

Is worth less than:

$1p \leftarrow 2p \leftarrow 5p$

Is worth more than:

$50p \rightarrow 10p \rightarrow 5p \rightarrow 2p \rightarrow 1p$

Adding money by touch

Make up these amounts of money by touch alone.

$3p = 2p + 1p$ or $1p + 1p + 1p$

$5p = 2p + 2p + 1p$ or $2p + 1p + 1p + 1p$
 or $1p + 1p + 1p + 1p + 1p$

$7p = 2p + 2p + 2p + 1p$ or $2p + 2p + 1p + 1p + 1p$
 or $2p + 1p + 1p + 1p + 1p + 1p$
 or $1p + 1p + 1p + 1p + 1p + 1p + 1p$

Try other sums of money.

Play 'Speed money'

Who can sort 1p and 2p coins most quickly on to a set grid?

Play money dominoes

Mark up your dominoes from card. Make sure that each written value has a corresponding value drawn in coins.

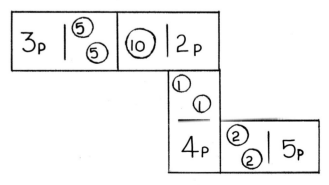

Shopping

Use coins to make up values for the purchases. Encourage children to make purchases for 10p, 20p, 50p etc.

Or they could spend: 10p to make 3 purchases
20p to make 7 purchases
50p to make 10 purchases.

Toy shop

Florists

Tuck shop

Collect old chocolate wrappers and sweet wrappers and put them around Plasticine to make your stock. Also use marbles.

Stationers

Use greetings cards designed by the children.

It is useful to make a collection of three-dimensional shapes. Often these can be displayed in a shop.

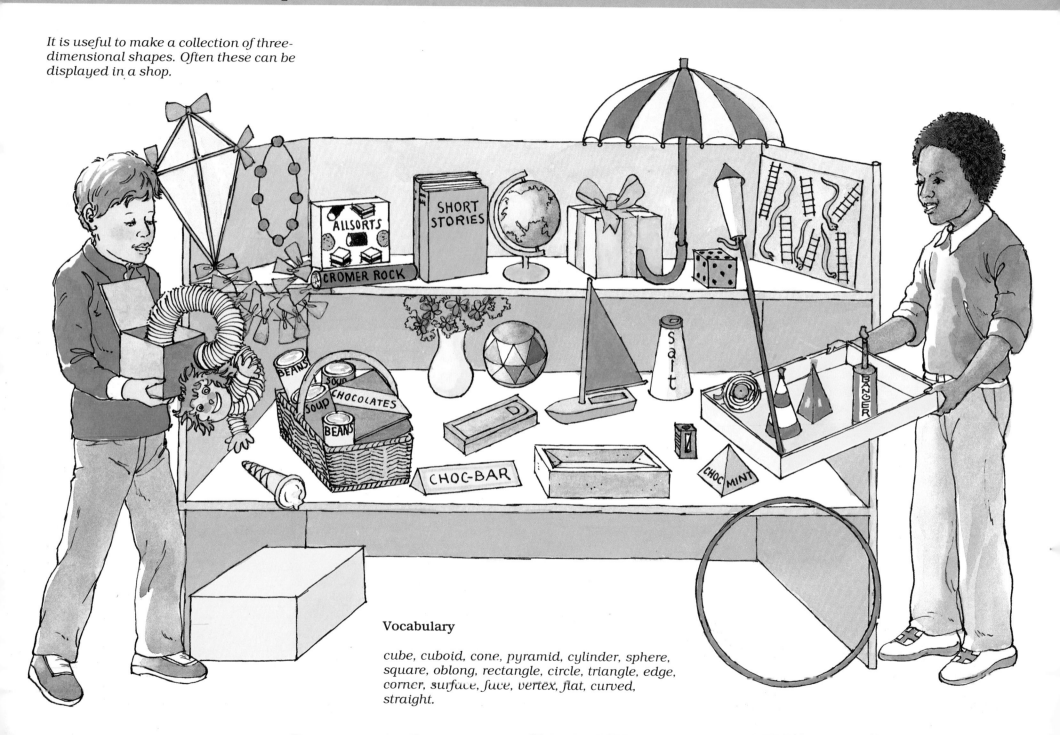

Vocabulary

cube, cuboid, cone, pyramid, cylinder, sphere, square, oblong, rectangle, circle, triangle, edge, corner, surface, face, vertex, flat, curved, straight.

Junk modelling

Like building with blocks, this involves the handling of three-dimensional shapes and a growing familiarity with prisms, pyramids and other shapes. Experience of stability and the fitting together of surfaces also come into play, as does the idea of area as the amount of surface covered.

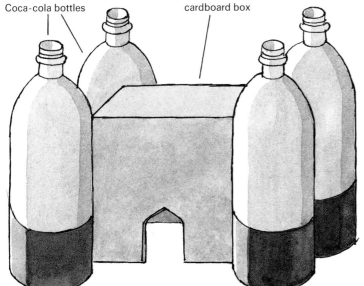

shoe box

chocolate box

toilet roll centre

Coca-cola bottles

cardboard box

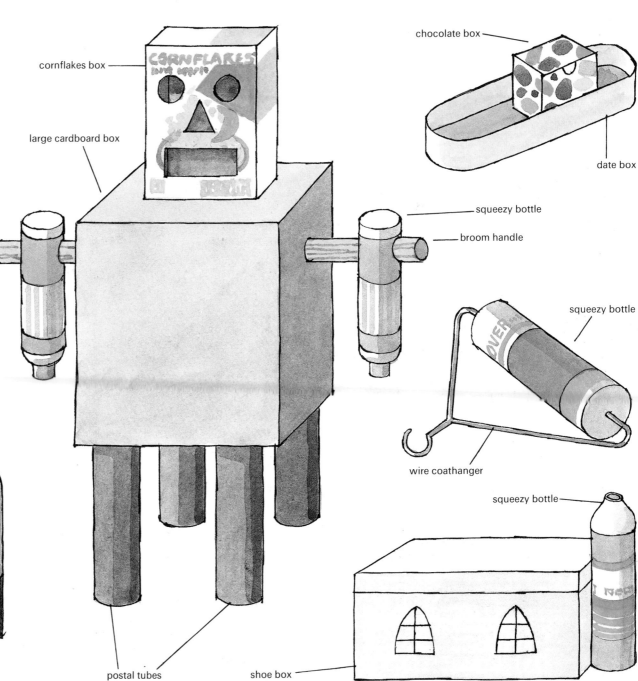

cornflakes box

large cardboard box

chocolate box

date box

squeezy bottle

broom handle

squeezy bottle

wire coathanger

squeezy bottle

postal tubes

shoe box

Collect bricks and blocks

These will help to develop the idea of form, pattern and space and show mathematical relationships.

wooden bricks

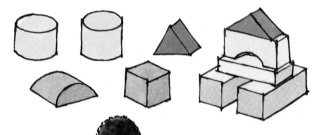

Hestair Hope's Jumbo Bag

Plastibrics

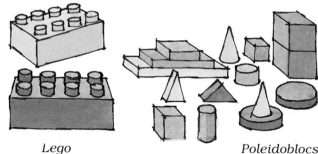

Lego

Poleidoblocs

Sorting by material

plastic

 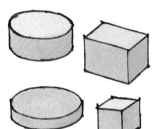

wood

clay (real bricks)

Sorting by size

small

large

thick

thin

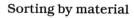

tall

short

Sorting by colour

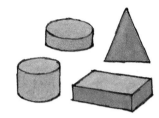

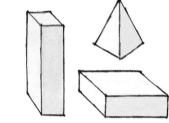

Sorting by shape

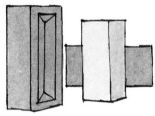 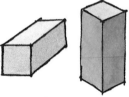 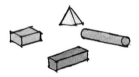

cube

cuboid

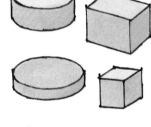 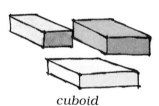

triangular prism

pyramid

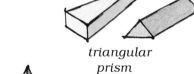 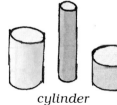

cylinder

Things which roll/things which don't

Solid things/hollow things

Flat surfaces/curved surfaces

Flat and curved surfaces

With vertices/without vertices

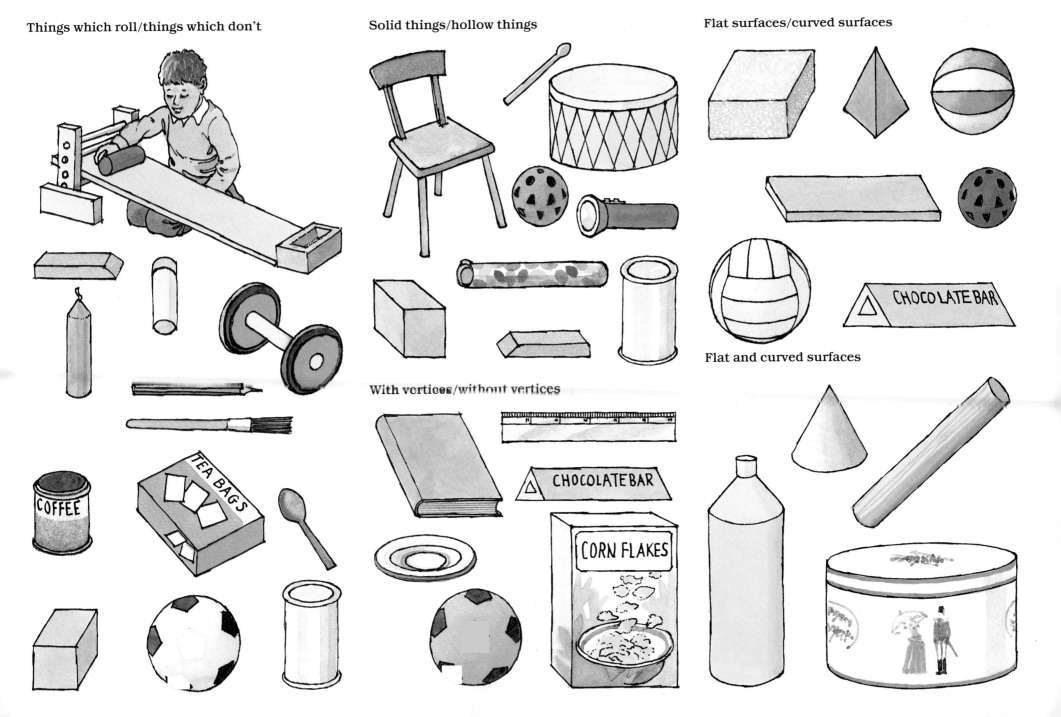

COFFEE

TEA BAGS

CHOCOLATE BAR

CHOCOLATE BAR

CORN FLAKES

Concentrate on these shapes to begin with.

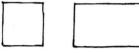

square   oblong   circle   triangle

Make a book of circles

Use coloured tissues.

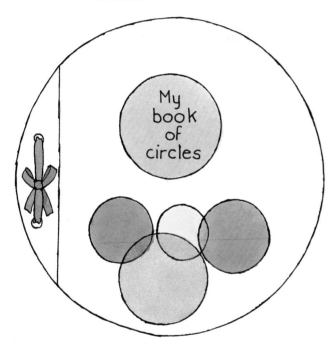

Make other books

Use templates

Use the following as templates and colour in the shapes.

Playing with shapes

Draw around a template to make pictures.

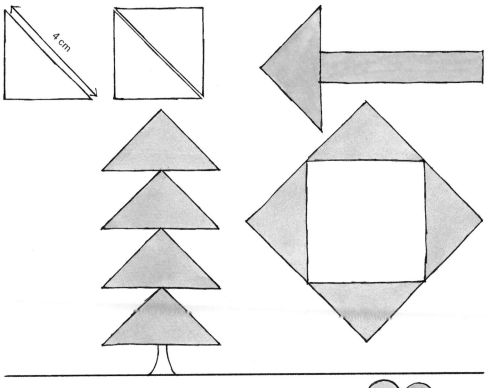

Make composite pictures

Use a mixture of templates to make composite pictures.

Look for patterns

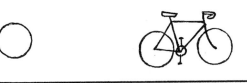

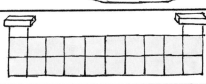

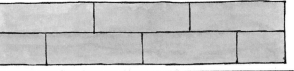

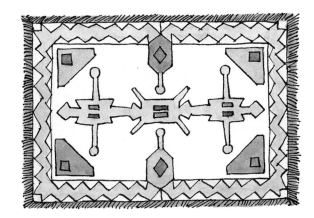

Any child's drawing of a house is usually symmetrical and children seem to have a feel for balance in their painting and constructional work.

Here are some activities that help to develop the concept of symmetry.

Blot pictures

Make some ink blots.

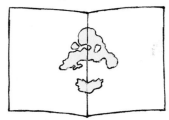

Fold and rub.

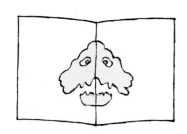

Elaborate to make a book of blot monsters.

Cut-outs

Take an A4 sheet.

Fold

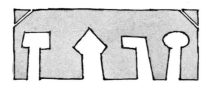

Cut

Open out and mount.

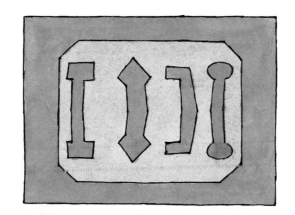

Pinprick pictures

Draw on a folded A4 sheet right up to the fold.

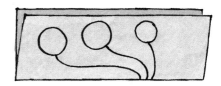

Prick through with a pin.

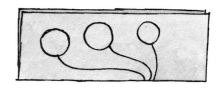

Open and draw through the pin holes.

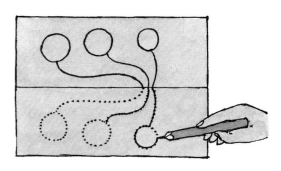

Symmetrical names

thick sugar paper
thick powder paint

Fold and press, then open out.

Repeating patterns

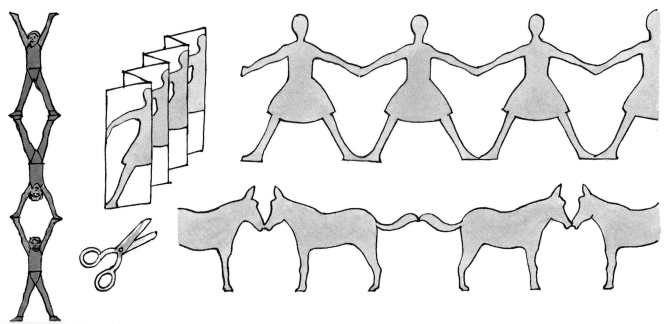

Finish the pattern

Make some pictures for children to complete the symmetry.

Make some mirror cards

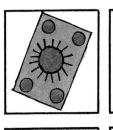

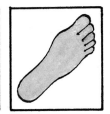

Hold the mirror so that you: extend the shape
make new shapes
complete a new
pattern
and so on.

Translation

A translation is merely a 'sliding along' of a shape in one direction.

The position of the shape is changed but its area remains the same.

Use cut-out shapes

Let the children cut out some shapes and slide them about.

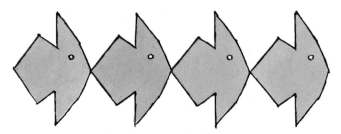

Doing this can lead to tessellation. A regular tessellation is one in which one shape can be used to cover a surface without any overlap or gap. A semi-regular tessellation uses more than one shape but in a repeating pattern so that the same shapes always meet at a vertex.

Make some tessellations

regular tessellation

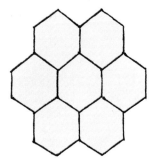

semi-regular tessellation

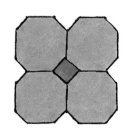

Tessellation introduces the concepts of area and angle.

With gummed paper shapes, you could cover an area.

Look through books for pictures of Roman and Byzantine mosaics and other tiling patterns.

Using squared paper, you could draw some tiling patterns.

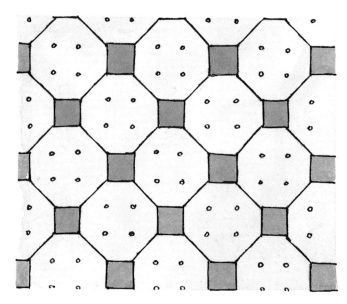

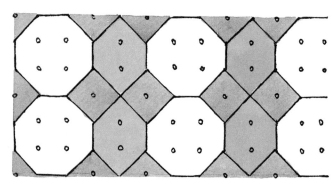

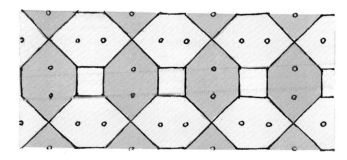

Some fun templates

Look outside for tessellating shapes

Make drawings and rubbings.

hanging tiles

slates on a roof

patterns in brickwork

stretcher bond

Flemish bond

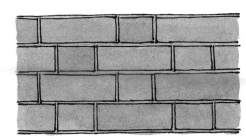

floor tiling

Prisms

A prism is a three-dimensional shape where the cross-section is congruent to the shape at each end.

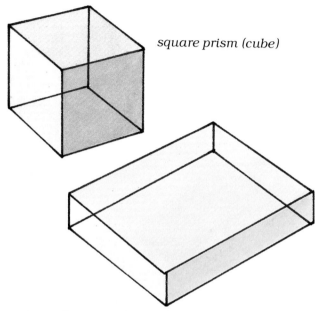

square prism (cube)

rectangular prism (cuboid)

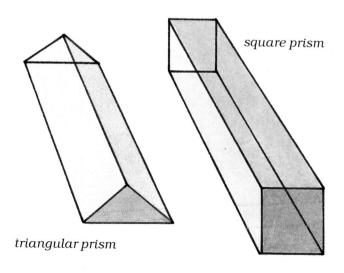

triangular prism

square prism

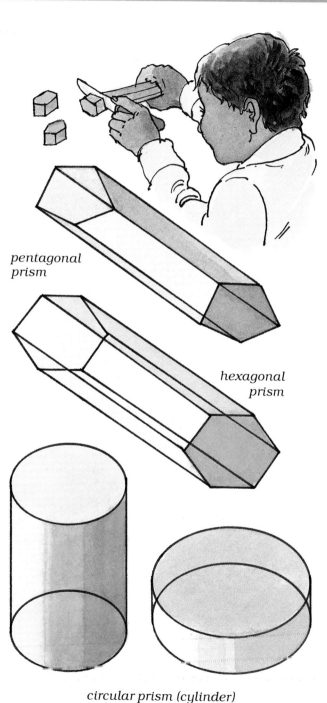

pentagonal prism

hexagonal prism

circular prism (cylinder)

Pyramids

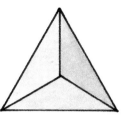

triangular pyramid (tetrahedron)

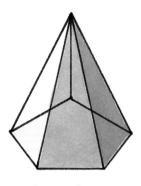

pentagonal pyramid

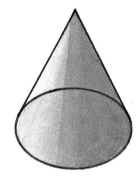

circular pyramid (cone)

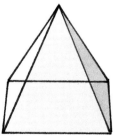

square-based pyramid

Others

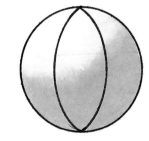

sphere

Two-dimensional shapes are often called plane shapes.

Triangles

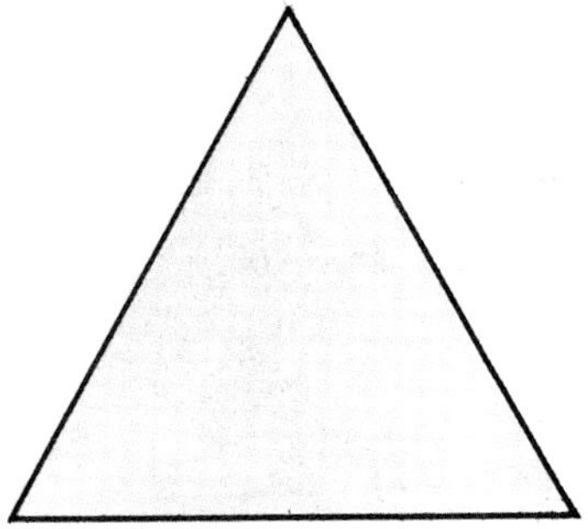

equilateral
(all sides equal)

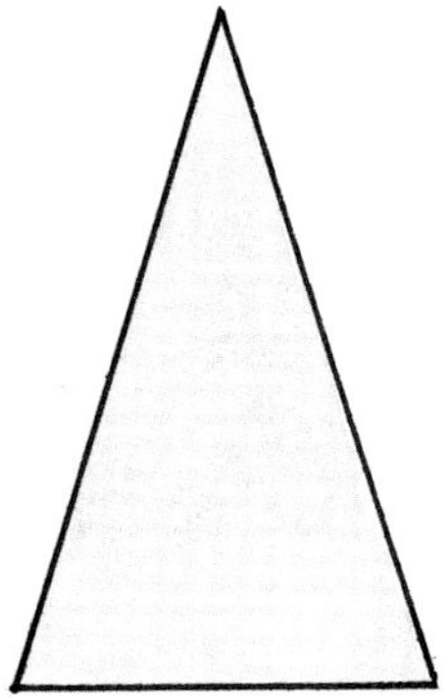

isosceles
(any two sides equal)

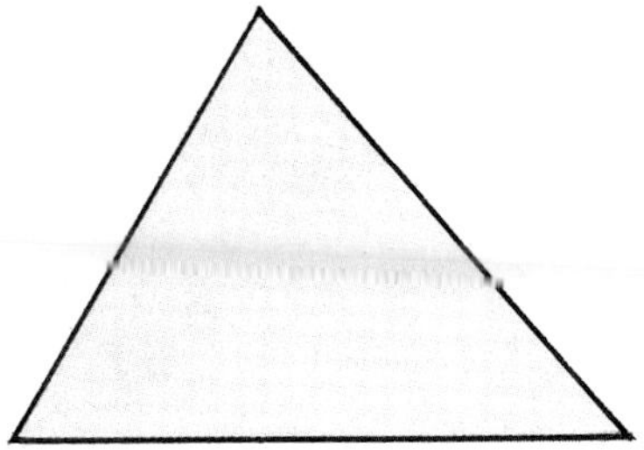

scalene
(all three sides different)

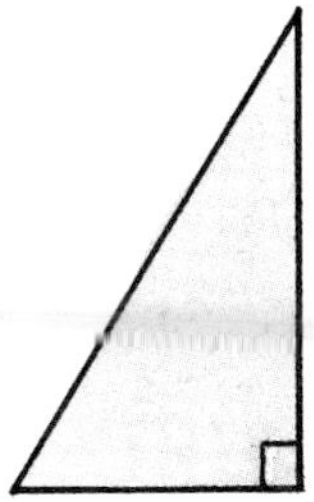

triangle with a
right angle

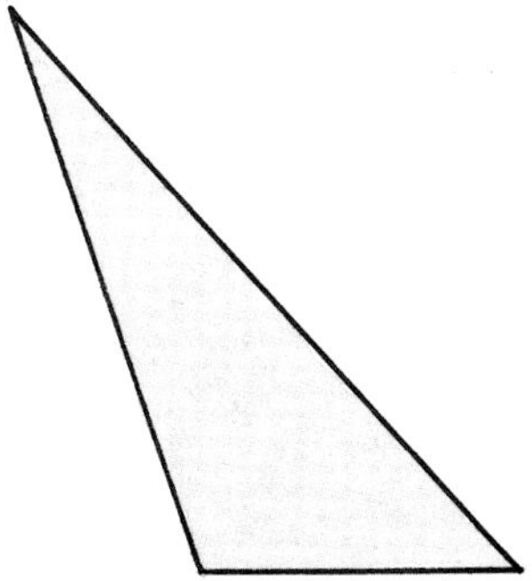

triangle with an
obtuse angle

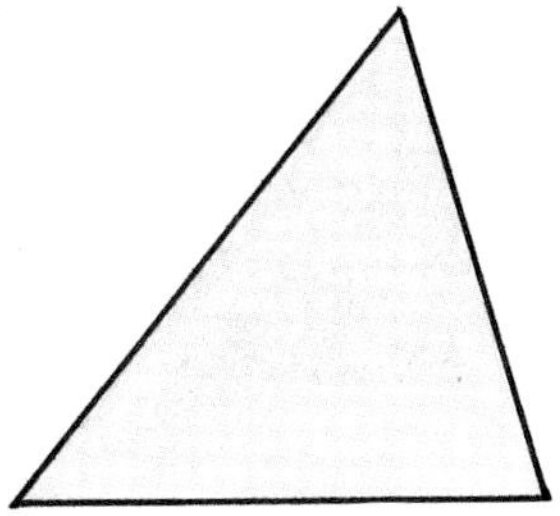

triangle with
acute angles

Quadrilaterals

square

oblong

parallelogram

trapezium

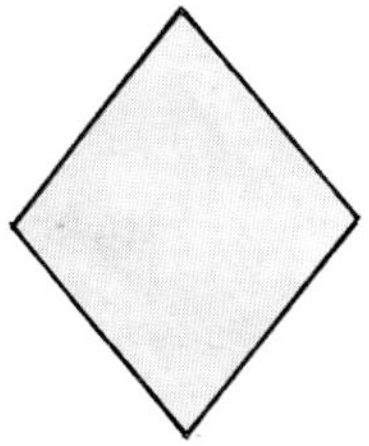

rhombus

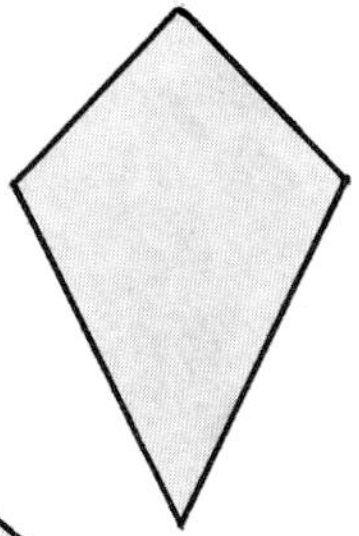

kite

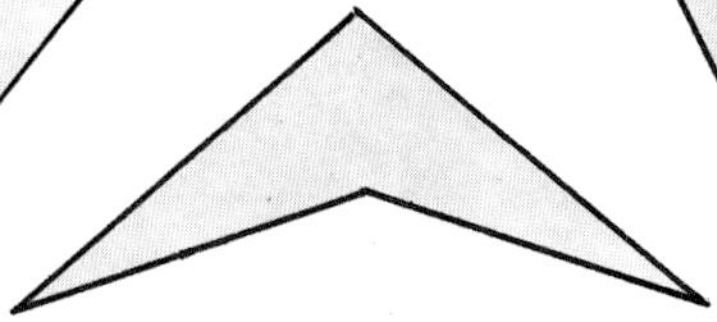

delta or arrowhead

Squares and oblongs are both rectangles.

Other polygons

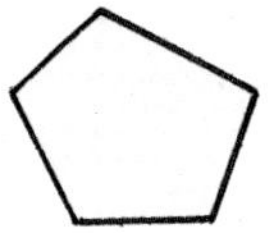

pentagon

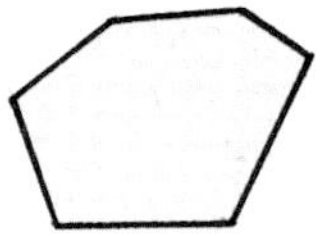

hexagon

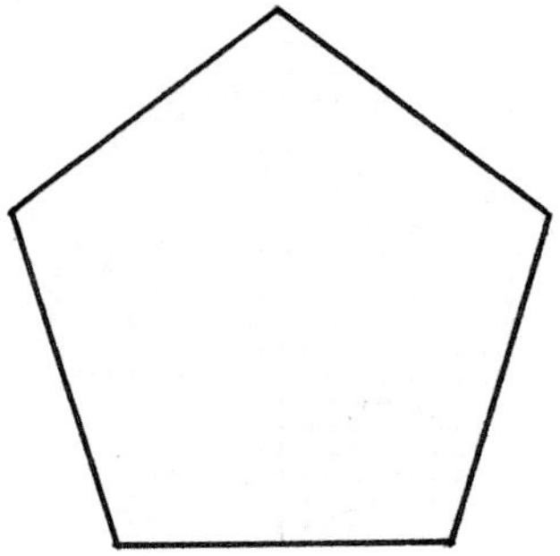

regular pentagon

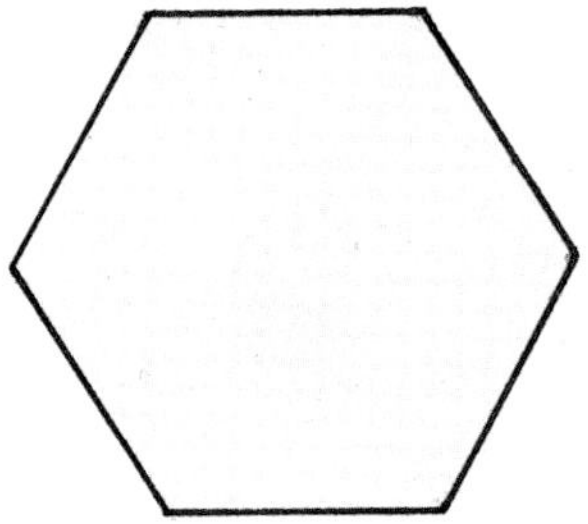

regular hexagon

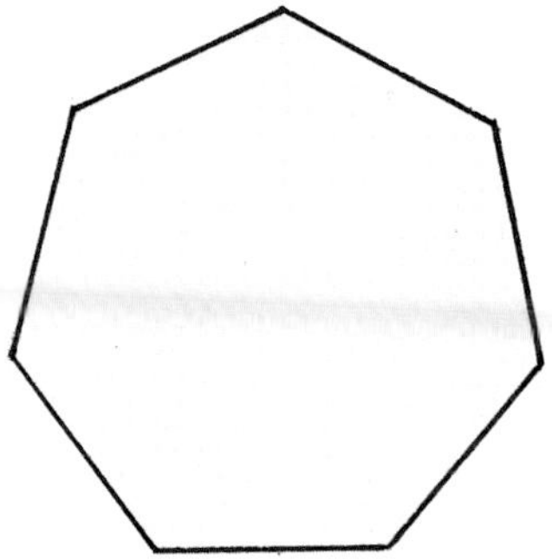

regular heptagon

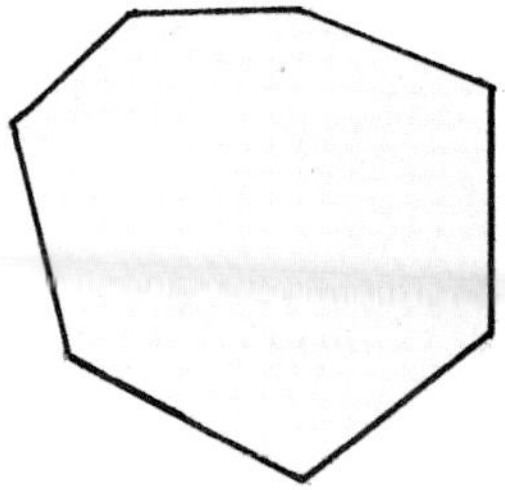

heptagon

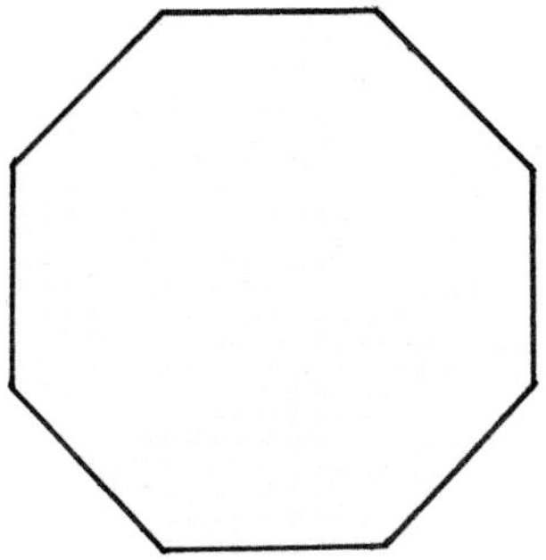

regular octagon

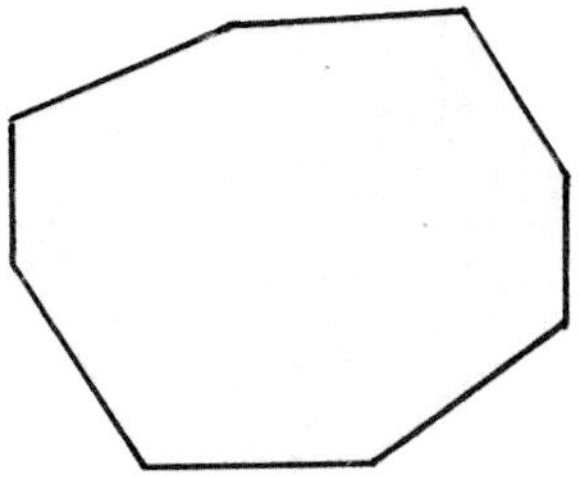

octagon

Use the simplest form of one-to-one correspondence to begin.

Children can use blocks to represent themselves.

All those with red jumpers stand on the mat.

All those without red jumpers stand on the floor.

Jasmindev
Rose
Barry
Hameed
Sheryl
Ramala
Tony

Wearing red jumpers

Not wearing red jumpers

Duplo people with red tops.

Duplo people without red tops.

Pictorial representation should provide a simple way of displaying information.

Where possible, children should discuss and choose the type of representation used.

It is valuable to display the same information in a variety of ways.

When the data is displayed, children should be encouraged to discuss and answer questions about the pictorial representation.

Proceed to squares which are stuck on.
This will lead towards block graphs.

Which school dinner?

Favourite toys

				Computers
				My Little Pony
				Bikes
				Lego
				Skateboards

How many conkers balance the parcels?

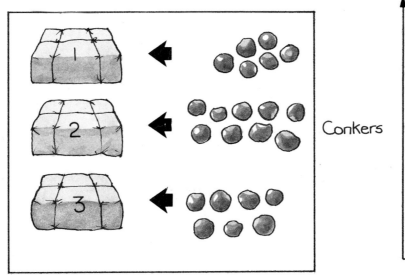

Conkers

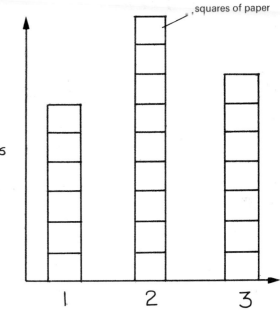

squares of paper

Graphs which make themselves

How many beads can you thread in a minute?

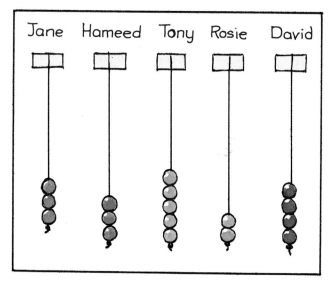

How many blocks can you stack in a minute?

Zahour | John | Julie

Which is the tallest toy?

Big Teddy | Grey Teddy | Rag Doll | Batman

Who has the longest arm?

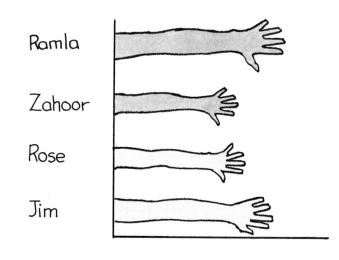

Ramla

Zahoor

Rose

Jim

Make and use a shadow stick

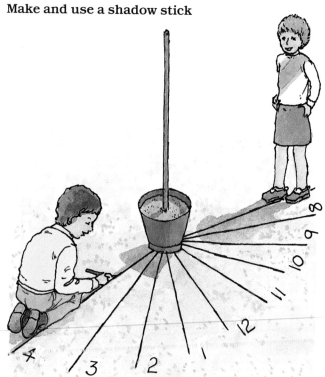

Tallies

Tallying can be visually very clear if set out in its original form.

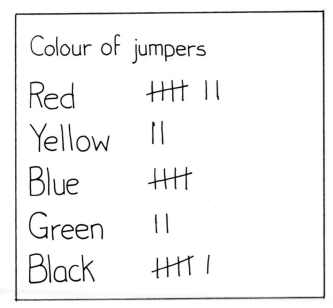

Colour of jumpers

Red	ⵜⵜⵜ II
Yellow	II
Blue	ⵜⵜⵜ
Green	II
Black	ⵜⵜⵜ I

Block charts

Here each object is represented by a square or rectangle, i.e. a block.

Birds in the playground.

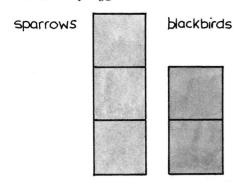

sparrows   blackbirds

Always keep the blocks the same size.

Bar charts

The height of upright rectangles (bars) can represent numerical data. Always mark in the vertical and horizontal axes.

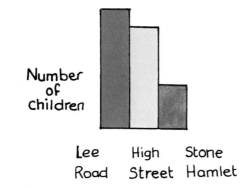

Number of children

Lee Road   High Street   Stone Hamlet

Bar lines

Here lines are used instead of bars. The arrow on an axis shows that the number could be continued if necessary.

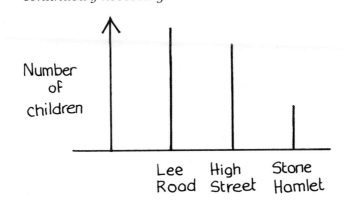

Number of children

Lee Road   High Street   Stone Hamlet

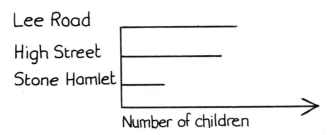

Lee Road

High Street

Stone Hamlet

Number of children

Some things for pictorial representation

Ourselves:

height	number in class
mass	boys/girls
shoe size	birthdays
length of foot	number in family
colour of eyes	bedtimes

Favourite things:

television programmes	pets
sweets	colours
pop groups	toys
story books	drinks
things to wear	biscuits

Other things:

make of staff cars
colour of staff cars
number of male/female staff
number of trees in school grounds
number of windows/doors in school